FREQUENTLY ASKED QUESTIONS
ON
PUBLIC PROCUREMENT

A Reference Guide to Procurement
and Contract Administration Basics

JORGE A. LYNCH T.

PROCUREMENT CLASSROOM SERIES, VOLUME #3

Also by Jorge A. Lynch T.

Public Procurement and Contract Administration:
A Brief Introduction

Public Procurement:
Principles, Categories and Methods

What is Public Procurement?

Essential Principles of Public Procurement

Procurement Planning Basics

Public Procurement Methods:
Identification and Selection

This book is intended to serve as a quick reference source for beginners and intermediate-level procurement professionals who wish to enhance their knowledge and build a successful career in Public Procurement. Material in this book is for educational purposes only. This book makes no guarantees of success or implied promises.

The Procurement ClassRoom believes in repurposing content whenever relevant and beneficial to our readers. Some materials in this book may have previously existed as a blog post, book excerpt, or another form of digital media.

About The Procurement ClassRoom Series

The Procurement ClassRoom was founded in 2013 to provide aspiring and novice procurement practitioners with a platform where they can interact and learn about the theory and practice of public and project procurement in a manner that is straightforward and easy to understand. Most importantly, with these fundamentals, they will be able to develop a solid foundation leading to a successful career.

This Guide is the third of the Procurement ClassRoom Series. It was developed primarily from the author's experience working on donor-funded projects in various countries since the year 2000. It relies heavily on donor and country-specific procurement guidelines that were developed based on the UNCITRAL Model Law on Public Procurement, but does not follow any of them to the letter; relying mostly on the author's practical experience and understanding. So, there are no specific references to any sources.

Other guides are in the making and will also be concise; containing sufficient information to give you a better understanding of different aspects of the management and practice of public and project procurement, and contract administration.

We appreciate receiving feedback from readers. Please send your suggestions for new topics and improvements to: suggestions@ProcurementClassRoom.com.

FREQUENTLY ASKED QUESTIONS
ON
PUBLIC PROCUREMENT

A Reference Guide to Procurement
and Contract Administration Basics

CONTENTS

FOREWORD

The body of knowledge known as Procurement is still in its evolutionary stage. As such, there is not much literature compared to traditional disciplines. Reading the "Frequently Asked Questions on Public Procurement: A Reference Guide to the Procurement and Contract Administration Basics" creates a scene of interaction between a lecturer and a student, procurement specialists and management, and a professional providing explanation to a curious member of the tax paying public. The Guide, in my opinion, meets the need of students, beginning procurement professionals, management, bidders and the curious public.

Many scholarly papers and textbooks discuss the subject of Public Procurement in complex language, thus creating difficulties and do not present a clear picture of procurement in practice or explain concepts in plain language. This guide presents answers, in a simplified manner, to questions asked by stakeholders and those curious of the processes involved in procuring public goods and services.

For students at the introductory and intermediate levels, it is a tool for study. Here, students can find definitions and explanation of concepts, that seem abstract, presented in plain language.

For the beginning procurement practitioner, it is a tool for quick reference to daily questions you may need to answer in your professional engagements.

For management, it is a book with the needed answers to questions forming in their minds daily. To them, it is also a means to getting familiar with the jargon they are confronted with as they make contracting decisions.

For bidders, it is a resource that can be referenced to understand needed requirements and answer questions, such is "what is a bid security", that they may be confronted with when preparing bids.

For the curious tax paying public who wants to know what is happening in the procurement process, it is a source for answers where you may not have a professional to refer to.

As you read through this rich Guide, you will find it interestingly a handy tool to assist you in your interactions in the world of the Public and Project Procurement practitioners. I am hopeful that your time spent reading the pages of this Guide would not be a waste. Enjoy your substitute specialist procurement practitioner and tutor in answering your frequently asked questions.

Aaron K.D. Cholopray
Procurement Manager, MCA-Liberia
April 2019

INTRODUCTION

When I started nearly three decades ago in procurement and related field, there were many questions I had while learning, and although some of them were answered by more senior professionals, many were left unanswered until I was eventually able to find the answers myself. So I often wondered what it would have been like to have had a manual that answered some of the basic questions beginners or aspiring practitioners may have on the various stages of the procurement process. This is such a text. It was developed with someone like you in mind. Someone eager to learn and would rather a text that got straight to-the-point in answering some of the most common questions in the field.

What is this book about?

This book is about Public Procurement. It is written in question and answer format, grouped within specific steps in the procurement process. It is intended to serve as a reference. Each chapter covers a stage in the procurement process. This is designed to help you find a quick and concise response to most questions you may have on the procurement process.

Who is this book for?

If you are new to the procurement profession, a student, or just interested in learning the fundamentals of public procurement, this book is for you. Intermediate level procurement practitioners may also find topics of interest in this book.

What will you learn from this book?

Learning and understanding the fundamentals of public procurement is important to your professional success. In this book you will quickly find answers to your questions on the different stages of public procurement and contract administration.

The following are the chapters covering specific topics:

1. Public Procurement and the Public Procurement Process,
2. Public Procurement Principles and Categories,
3. Public Procurement Planning and Scheduling,
4. Public Procurement Selection Methods,
5. Requesting Bids and Proposals,
6. Receiving and Opening Bids and Proposals,
7. Evaluating Bids and Proposals,
8. Contract Negotiations and Award,
9. Contract Administration Basics,
10. Contract Implementation Instruments,
11. Contract Monitoring,
12. Payments Processing, and
13. Contract Modification, Suspension and Conclusion.

There is also an appendix clarifying some of the common terms used in the book.

· CHAPTER 1 ·

PUBLIC PROCUREMENT AND THE PUBLIC PROCUREMENT PROCESS

1. What is public procurement?

Public procurement is the acquisition of goods, services and construction works to support government operations, and to provide public goods and services. Such services include: public transportation, utilities services, medical and educational facilities and services, roads, bridges and other public infrastructure, all of which are classified as either "hard infrastructure" or "soft infrastructure."

2. What is the public procurement process?

The public procurement process comprises all the steps taken to obtain goods services and works with public funds. It begins with the assessment of what the entity needs and ends with signing a contract with a supplier, contractor or service provider.

The following are the steps in the public procurement process:

1. Assess needs
2. Identify requirements
3. Do market research
4. Allocate budget
5. Select procurement method
6. Develop procurement plan
7. Prepare scope of work or specifications for Goods, Services or Works

8. Prepare solicitation/bidding documents
9. Prepare and advertise procurement notice
10. Conduct pre-bid meeting and site visit
11. Accept and respond to bidders' queries
12. Receive and open bids/proposals
13. Evaluate bids/proposals
14. Award contract

3. What is the goal of public procurement?

The goal of public procurement is to award contracts in a timely and cost-effective manner to qualified contractors, suppliers and service providers. Contracts are awarded for goods, services and works, to support government operations and provide public goods and services in accordance with the procurement rules.

4. Who are the actors in public procurement?

Procurement practitioners are the primary actors in the public procurement process. They are responsible for ensuring the goal of public procurement is achieved, must strive to gain stakeholder's trust and ensure full understanding of the procurement process. Procurement practitioners are engaged from needs assessment to contract close-out. They also support contract administration, although they are not solely responsible for this process.

5. Who are the Stakeholders in public procurement?

Any person, group or organization that can affect or be affected by the public procurement system is a stakeholder. Some examples of stakeholders in public procurement are the inhabitants of a country, businesses, government institutions, bilateral and multilateral donors and other institutions, public procurement professionals and other civil servants.

6. Who are the beneficiaries of public procurement?

All inhabitants of a country, citizens or not, benefit from the public procurement system because of the public goods and services available and provided to them in the form of transportation systems, public utilities, public education, and medical services and facilities, to mention a few.

7. What governs the public procurement process?

The public procurement process is governed by the Public Procurement Rules. These rules are composed of the legal and regulatory frameworks that must be followed. The Public Procurement Rules often include a procurement act, regulations, guidelines, manuals, standard solicitation (bidding/tender) documents and contract forms. These may address everything from determining, planning and scheduling what to buy, when and where to buy, in addition to contract implementation and management.

8. What is contract administration?

Contract award marks the beginning of the contract administration phase, which deals primarily with the management of risks in contract implementation. The main purpose of contract administration is to monitor performance to ensure the objectives of the contract are met on time and within budget, and to detect deficiencies and find ways to resolve them without affecting the expected outcome of the contract.

Contract administration is often neglected because of poor planning. In the acquisition of goods, services and works, the focus is mainly placed on the procurement process (identification of the requirement up to contract award); although this is often a comparatively shorter period than contract implementation. It is also only after a contract is awarded that the real benefits of the procurement process are realized; so more attention should be placed on ensuring the contract is implemented as intended and agreed upon.

· CHAPTER 2 ·

PUBLIC PROCUREMENT PRINCIPLES AND CATEGORIES

9. What are public procurement principles?

Public procurement principles are the foundation of public procurement and should be the basis for the public procurement rules. Principles govern the management of public procurement and set the framework for a code of conduct for public procurement practitioners and others directly or indirectly involved in the public procurement process.

Practitioners should have a clear understanding of public procurement principles, knowing how to use them to guide their day-to-day decisions. By integrating these principles into their work ethic, the outcome of their decisions will be in line with the goal of public procurement.

10. What are the seven fundamental principles of public procurement?

The seven primary principles of public procurement are:

1. transparency,
2. integrity,
3. economy,
4. openness,
5. fairness,
6. competition and
7. accountability.

10.1 What is transparency?

The purpose of transparency is to ensure all information and documents on the public procurement process are available to and shared with all stakeholders: contractors, suppliers, and service providers, unless there are valid and legal reasons for keeping certain information confidential. When a public procurement requirement is advertised, the announcement must include enough details for interested contractors, suppliers and service providers to assess if they are qualified and interested in submitting a bid or proposal. The solicitation documents must be available at a reasonable price; preferably, free of charge. After reading the solicitation documents, interested bidders should be able to know:

1. the nature of the requirement and its scope,
2. the evaluation and selection criteria,
3. how and where bids or proposals must be submitted,
4. the number of copies, and point of contact for additional information and response to queries (clarifications),
5. the deadline to request clarification of the solicitation documents,
6. the schedule of pre-bid meetings and site visits, and
7. the closing date for submission of bids, proposals or information,
8. other pertinent details.

If there are changes to the solicitation documents, all bidders should be notified through the same newspapers, websites or publications that were used for the initial notification.

10.2 What is integrity?

Integrity is reliability. Bidders, and other stakeholders, must be able to rely on any information circulated by the procuring entity, formally or informally. In public procurement integrity is two-fold; there is the integrity of the procurement process, and the integrity of the public procurement professional.

10.2.1 What is integrity of the procurement process?

The integrity of the procurement process assures confidence in the public procurement system. When documents are publicly available, the information they contain must be dependable and free of ambiguities or bias.

When reviewing solicitation documents, prospective bidders should be able to determine if they are qualified to undertake the assignment. If they are not qualified, they should be able to assess the need for association with other bidders, in the form of joint ventures or subcontracting arrangements.

Bidders should have a clear understanding of what is required and how they will be evaluated. The evaluation and selection criteria must be clearly stated in the solicitation documents. These criteria should remain unchanged. If modification is required, the solicitation documents should be amended, published and made available to all prospective bidders. Any changes in the bid submission date should allow bidders enough time to adjust their bids accordingly to meet the new submission deadline.

10.2.2 What is integrity of the public procurement professional?

Integrity of public procurement professionals has to do with the public's perception of these public servants in the execution of their duties. Practitioners, such as procurement officers or specialists, and other government officials involved in the public procurement process, must demonstrate personal and professional integrity. Ideally, there should not be any contradiction between personal and professional integrity.

Public servants involved in the public procurement and contract administration process should always be perceived as honest, trustworthy, responsible and reliable. They must keep the purpose of the procurement requirement in mind and strive to ensure they responsibly manage public procurement and contract administration in accordance with the public procurement rules.

10.3 What is economy in public procurement?

This principle is synonymous with efficiency and value for money. Economy emphasizes the need to manage public funds with care and due diligence so that prices paid for goods and services are acceptable and represent good value for the public funds expended on them.

Everyone associated with the public procurement process or directly responsible for enabling the acquisition of goods and services with public funds, should strive to avoid fraud, waste and abuse of public resources; whether it is the result of over specification of required goods, paying unreasonably high prices for substandard goods, collusion with or between bidders, or other forms of unacceptable practices.

10.4 What is openness in public procurement?

Openness means public procurement requirements should be available to all qualified organizations and individuals. The public should have access to information about public procurement requirements. However, access to public procurement information is not absolute, because confidential and proprietary information of organizations and individuals participating in the procurement process should not be publicly available without their approval.

With respect to procurement methods, such as restricted/ selective bidding, they limit the availability of solicitation documents to firms meeting certain qualifications. The request for quotations (or shopping), and direct contracting (sole source) are also procurement methods that limit competition, because the invitation for bids is issued only to a limited number or only one organization or individual.

The bid evaluation process is always confidential; although the results should be publicized during the notification of intent to award.

The details of defense procurements are usually kept confidential, restricting relevant information to a "need-to-know" basis. Except for confidential defense procurements, the results of the public procurement process should be published and made available on relevant websites.

10.5 What is fairness in public procurement?

Fairness in public procurement pertains to ensuring all bidders are treated equally and without bias. The following are important points towards exercising fairness in the public procurement process:

1. Decision-making must be impartial. No special treatment should be extended to individuals or organizations because public procurement activities are undertaken with public funds.
2. All bids must be considered based on their compliance with the stipulations of the solicitation documents, and bids should not be rejected for reasons other than those stated in the solicitation documents and public procurement rules.
3. A contract should only be signed with the supplier, contractor or service provider whose offer is compliant, and best responds to the objectives of the requirement in terms of technical capability and price.
4. Suppliers, contractors or service providers should have the right to challenge the procurement process if they believe they were unfairly treated or that the procuring entity failed to carry out the procurement process in accordance with the public procurement rules. Such challenges must be based on the conditions specified in the solicitation documents and the public procurement rules.

10.6 What is competition in public procurement?

Competition in public procurement involves inviting a reasonable number of companies or individuals to submit offers for procurement requirements for which they qualify. One of its principal aim is to award competitively priced contracts based on the best solution offered that is compliant with the specifications and scope of work of the requirement.

The public procurement process should not be manipulated for the benefit of any organization or individual. As public procurement is funded primarily with taxpayers' money, all eligible organizations and individuals should be allowed to participate by submitting bids in response to a specific requirement for which they are qualified.

Public procurement requirements should be widely disseminated to increase the chances of a good market response, leading to the award of competitively-priced contracts.

Despite this principle, not all contracts are awarded using a competitive process because the procurement method used may depend on the urgency of need or other conditions. The use of non-competitive procurement methods, although justified under certain conditions, should be kept to a minimum.

10.7 What is accountability in public procurement?

Accountability in public procurement is the responsibility procurement practitioners have to report on and explain their actions and decisions made throughout the public procurement process. As public servants, procurement practitioners, and others involved in public procurement and contract administration, may be held accountable and exposed to sanctions for any behavior that contravenes the public procurement rules.

11. What is the framework for developing a public procurement code of conduct?

A Public Procurement Code of Conduct should be developed based on the fundamental principles of public procurement, which should be defined in the public procurement rules. A code of conduct serves the purpose of guiding procurement practitioners in what is acceptable and unacceptable behavior in the public procurement process. Practitioners are obliged to adhere to a relevant Public

Procurement Code of Conduct in their day-to-day procurement-related decisions.

12. What are the primary public procurement categories?

The primary public procurement categories are goods, services and works. These are the basic categories within which all procurements are divided. All procurement activities will correspond to one of these categories, and there are many subcategories stemming from each of them.

13. What are goods in public procurement?

Goods are physical products that may be purchased off-the-shelf or manufactured on request. Typical examples of goods are office supplies and equipment, furniture, IT equipment, books, vehicles, medical supplies and other commodities. There may be a service element to providing these goods, such as when they need to be assembled or installed, or both. But the extent of the service involved is directly related to the acceptance of the goods purchased.

14. What are services and how many types are there in public procurement?

Services are classified as either i) consulting services or ii) non-consulting services. In some cases, they are simply classified as services because of the difficulty, at times, in clearly determining the difference between the two. The distinguishing factor between them, however, is the degree of importance of the measurable physical output of the services provided.

14.1 What are consulting services in public procurement?

Consulting services usually do not require the extensive use of equipment to produce an output. These services are generally advisory in nature and project related. They include feasibility studies, project management services, technical assistance, training and capacity development, finance and accounting services, to mention a few.

14.2 What are non-consulting services in public procurement?

Non-consulting services typically involve the use of equipment and a specific methodology to achieve the desired results. Some common examples of non-consulting services are equipment maintenance and repair, operations and maintenance services, utilities management, installation and maintenance services, surveys and field investigations, and automotive maintenance and repair services, to mention a few.

15. What are works procurement?

Works are also called infrastructure works; they include the construction of all types of structures (buildings, highways, bridges, and similar structures), renovations, extensions, demolitions and repairs. This category also includes, water and sanitation, transportation and electrical plant infrastructure.

Goods and works procurements are usually provided by business organizations, while consulting and non-consulting services may be provided by businesses as well as individuals, depending on the uniqueness of the requirement.

· CHAPTER 3 ·

PUBLIC PROCUREMENT PLANNING AND SCHEDULING

16. What is procurement planning?

Procurement planning is the process of forecasting what to buy, for which requesting entities to buy, and when. To ensure procurement requirements are fulfilled with minimum delays, these projections should be made long before the goods and services are needed.

17. What are the steps in the procurement planning process?

There are ten steps in the procurement planning process:

1. Identify the requirements of each requesting entity.
2. Determine market availability (local, international or both).
3. Estimate the budget.
4. Seek approval to commit funds.
5. Select procurement method.
6. Identify similar procurements and package them to reduce the number of requirements and for economy of scales.
7. Prepare draft procurement plan and discuss with relevant requesting entities.
8. Complete the procurement plan.
9. Get procurement plan approved.
10. Prepare procurement schedule.

18. What is a procurement plan?

A procurement plan is a list of procurement requirements that are needed over a period (usually one year or more) to fulfill the needs of an annual budget or a project plan. Each requirement for goods, works or services is either to satisfy an immediate or future need. All requesting entities that are served by the procuring entity should prepare a preliminary procurement plan, with a brief description of their needs for goods, works and services, their budget estimate, and, preferably, a tentative date they expect the contract to be awarded.

19. When is the procurement plan prepared?

The procurement plan is usually prepared once a year and updated every six months. It depends on what is stipulated in the procurement rules. A procurement plan can also be amended to include unforeseen requirements; however, amendments should be kept to a minimum to encourage all departments to plan in order to reduce the need to expedite unforeseen procurements and put undue burden on the procuring entity.

In project or program procurement, a procurement plan is prepared at the beginning of the project or program and is then updated periodically, usually every six months, over the life of the project.

20. Who prepares the procurement plan?

Each requesting entity should prepare a preliminary procurement plan, with the assistance of the procuring entity. The procuring entity is responsible for preparing a consolidated procurement plan, which combines the requirements of all the entities supported by the procurement entity. In the case of a departmental procuring entity, the procurement plan will consist of all the procurement requirements of the units served by the departmental procuring entity.

21. What is a procurement schedule?

The procurement schedule is a detailed version of the procurement plan. Its purpose is to forecast how long it will take from initiating the selection process (after receipt of the terms of reference or technical specifications) until the contract is awarded. The procurement schedule lists all the milestones of each step in the selection process. Useful procurement performance information can be gathered from the procurement schedule by monitoring the planned versus actual completion of milestones. This information can alert the procuring entity to bottlenecks in the procurement process and inefficiencies in the procurement system.

22. What is the difference between a procurement plan and a procurement schedule?

The procurement plan is a list of all procurement requirements expected to be procured over a certain period, six months or more. This list includes a description of each requirement, the procurement category (goods, works, services), the procurement method, the allocated funds and the department for which the procurement is being carried out; sometimes an estimated contract award date is also included. The procurement schedule, in contrast, includes all the information that is on the procurement plan; but what differentiates the two is that the procurement schedule also includes all milestones and specific time periods for each step in the procurement process, from the preparation of the terms of reference or technical specifications up to the award of the contract. The procurement schedule may also be used to monitor the results of the procurement process for each requirement and the overall performance of the procuring entity.

· CHAPTER 4 ·

SELECTION METHODS
IN PUBLIC PROCUREMENT

23. What are procurement methods?

Procurement methods are the procedures used for selecting suppliers, contractors and service providers from which to obtain goods, works and services. Procurement methods may be competitive or non-competitive, as determined in the applicable procurement rules. Open tendering, restricted tendering, and single-source procurement are examples of procurement methods. The steps to follow in the selection process is determined by the procurement method; this has a direct impact on the time it will take from the initiation of the selection process until contract award.

24. How are procurement methods selected?

Procurement methods are selected based on the procurement category (goods, works or services); the allocated budget and complexity of the procurement requirement are also important. There are, generally, five factors to consider when selecting a procurement method:

1. **Procurement methods are stipulated in the procurement rules**. Procurement methods are not selected at random. All procurement methods are stipulated in the applicable procurement rules, and procurement practitioners must select the appropriate procurement method for each procurement requirement.

2. **The allocated budget and the complexity of the procurement requirement**. These are factors used to determine if the procurement method should be competitive or non-competitive. Usually, procurement requirements that are below a certain monetary threshold and are for readily available goods are assigned a non-competitive procurement method, such as shopping.

3. **The degree to which the procuring entity can clearly define the requirement**. There are instances where a requirement cannot be clearly defined and needs input from prospective bidders. Once this is determined, the most appropriate procurement method can be identified.

4. **The urgency of need**. This can result in the need to use direct or single-source procurement, especially in case of emergencies.

5. **Market availability**. Are there only a few firms or individuals with the expertise or goods needed to fulfill the requirement? This would be a strong justification for using a restricted or single-source procurement method. Market availability is also crucial to determining the need for international competitive bidding.

25. What are the different types of procurement methods?

Procurement methods are classified as competitive and non-competitive. Competitive procurement methods are those where the procurement notices are advertised and open to all qualified firms. Open tendering and two-stage tendering are examples of competitive procurement methods. Such methods are preferred because they presumably tend to promote transparency, economy and efficiency, and limit favoritism. Non-competitive procurement methods, in contrast, are seldom advertised, and they tend to limit the number or types of firms or individuals that may be invited to submit offers (bids or proposals). A few examples of non-competitive procurement methods are the request for quotations or shopping, restricted tendering, and single-source or direct procurement.

· CHAPTER 5 ·

REQUESTING BIDS AND PROPOSALS

26. What is a procurement requisition?

A procurement requisition is used to confirm the availability of funds before initiating the procurement process. To commence the procurement process for any procurement requirement listed on an approved procurement plan, a procurement requisition is prepared and sent to the procuring entity to prepare the solicitation documents which are then released to market to invite bids.

27. What information is needed on the procurement requisition?

The procurement requisition should identify what is being requested, the name of the requesting entity, the procurement plan reference number, the allocated budget, a certification of availability of funds, and the signature of the approving authority. Ideally, the terms of reference or technical specifications should be attached, clearly describing the details of the procurement requirement for goods, services or works.

28. What are solicitation documents?

Solicitation documents are the documents prepared and used by the procuring entity to invite bids and proposals for goods, services and works from prospective bidders. They are also called bidding documents or tender documents.

29. When are solicitation documents prepared?

Solicitation documents are prepared by the procuring entity after receiving an approved procurement requisition for any of the items that are on the approved procurement plan.

30. Who prepares the solicitation documents?

Solicitation documents are prepared by the procuring entity after receipt of an approved procurement requisition certifying availability of funds for any item on the approved procurement plan. They are prepared using the specifications, terms of reference or statement/scope of work, reviewed and agreed with the procuring entity and other relevant departments before submission of the procurement requisition.

Depending on the complexity of the procurement requirement, the specifications, terms of reference or statement/scope of work may be prepared by a technical expert hired for that purpose. In that event, the technical expert would prepare specifications, terms of reference or statement/scope of work and send them to the relevant departments, including the procurement entity, for review and approval. They may also prepare the solicitation documents, if that was agreed in their contract.

31. What are the common sections of the solicitation documents?

The most common sections of the solicitation documents are:

1. Invitation: A general overview of the requirement;
2. Instructions to Bidders or Consultants: Information on the bidding process, preparation, submission and evaluation of bids or proposals;
3. Data Sheets: Specific data amending or enhancing certain clauses of the instructions section. Gives details on date of submission of bids/proposals, and information about the purchaser (employer) and the bidder;
4. Qualifications: Information on requirements or criteria

bidders need to meet to qualify for the assignment;

5. Bid or Proposal Forms: The different forms bidders must use to submit their bids or proposals;

6. Evaluation Criteria: Criteria used to evaluate bids or proposals submitted, to determine to what extent they meet or exceed the requirements of the solicitation documents;

7. Specifications or Scope of Work (Terms of Reference): A description of the goods and/or services required;

8. Contract Template: Draft format of the contract that will be signed between the winning bidder and the purchaser or employer, and

9. Contract Conditions: Contains the general and specific (particular) conditions of the contract.

32. Who approves the solicitation documents?

The approval of solicitation documents follows the procurement rules and is based on the procurement method and monetary value. The procurement rules stipulate the approval requirements. Approval begins with the technical specification or terms of reference, which is fundamental and an integral part of the solicitation documents.

Government (Public) Procurement

In government procurement, the approval process usually involves a tender board. There may be several levels of tender boards which would have approving authority depending on the monetary value of the procurement. There may be a departmental tender board, ministerial tender board, and a cabinet tender board. Each of these have authority at different monetary value thresholds, depending on the procurement method.

The approval process is done in different stages and may involve different tender boards. For instance, if the highest approval authority for a specific procurement method is ministerial, the departmental tender board would do the initial approval before the document is sent to the ministerial tender board. In this case there would be two approvals. But in cases where the procurement method require cabinet approval, the departmental tender board would initially approve, then the ministerial tender board, before the documents are sent for the final approval of the cabinet tender board. But, if the procurement method calls for only a departmental approval, after the departmental tender board approves, there is no need for the approval of any other tender board.

Donor-Funded Project Procurements

In the case of donor-funded projects, where donor guidelines (such as those of the development banks) are used, it needs to be determined which procurement requirements need prior review or post review of the donor entity. Despite this, there may be (and usually is) a need for the approval by certain government entities, such as tender boards, or contracts committee either before or after the donor entity's approval. This is determined based on the agreement with the donor entity; because although donor funds are used, the procurement process may still need to follow certain stipulations of the government procurement rules for internal processing of the solicitation documents and other procurement actions.

If the procurement activity must undergo prior review of the donor entity, the solicitation documents need the approval (or no objection) of the donor entity before the invitation or call for proposals is advertised. If post review, such "no objection" is not needed, but the donor entity has the right to periodically review the procurement records to determine if the process was carried out in accordance with the loan or grant agreement, and the donor may claim/invoke mis-procurement if they consider that the agreed procedures were not followed for certain procurement activities.

33. When are procurement notices published?

Publication or advertisement of procurement notices depend on the procurement method. Competitive Bidding procurements are usually advertised in the local and sometimes international market.

But non-competitive bidding procurements such as request for quotations, direct contracting and sole-source procurements are not advertised.

Before publication, after receipt of an approved procurement requisition, the solicitation documents are prepared and approvals obtained, if needed, and then the procurement notice is advertised in local newspapers and internationally when required. The notice is also posted on approved websites. In newspapers and other paid media, the procurement notice is published for several days, but this period should be indicated in the procurement manuals and guidelines.

34. What are pre-bid meetings and site visits, and what is their purpose?

Pre-bid meetings and pre-proposal meetings are scheduled for any procurement where there is a need to meet with prospective bidders, before the submission of their bids or proposals, to clarify any questions they may have on any aspect of the solicitation documents. It is called a pre-bid meeting, when the procurement activity is for goods, works or non-consultant services and a pre-proposal meeting when it is for consulting services procurements.

During these meetings, prospective bidders can ask questions in the presence of representatives from the procuring entity and the technical team that prepared the specifications, terms of reference, and even the drawings bill of quantities (BOQ's), when the procurement requirement is for infrastructure works. At this meeting, a written record is prepared and a formal response, in the form of minutes, is sent to all prospective bidders that registered or purchased the solicitation documents. The formal response is binding on the procuring entity.

Site visits are conducted or permitted when necessary. They are done mostly for infrastructure works procurement, where prospective bidders are permitted to visit the site, either on their own or organized and conducted by the relevant entity in coordination with the procuring entity. When a pre-bid meeting is also scheduled, the site visit is held before the pre-bid to allow bidders to raise questions resulting from the site visit. These questions will be responded to during the pre-bid meeting and in the pre-bid meeting minutes. Only the formal (written) response is binding on the procuring entity.

· CHAPTER 6 ·

RECEIVING AND OPENING BIDS AND PROPOSALS

35. When are bids received and opened?

After the advertisement and release of the solicitation documents, bidders are required to submit their bids by a certain date and time indicated in the solicitation documents. Unless there is a formal extension of the submission date and time, bidders need to ensure their bids are submitted on or before the submission date. The opening of bids should take place on the date and time stipulated in the solicitation documents, which is usually on the same day and immediately after the submission time has passed.

36. How are bids received and opened?

Bids for goods, works and non-consulting services procurements are received and opened in a similar manner. Bids received before the submission date are stored until the opening date and time. The procuring entity is usually responsible for receiving and opening bids, but the Procurement Rules will indicate what entity is responsible for undertaking this function. Bids are received only up to the submission date and time indicated in the solicitation documents. After that date and time, bids should not be received, unless the submission date and time in the solicitation document were amended. If there is a bid box, it is closed on the submission date immediately after the submission time has passed.

The time for bid opening is indicated in the solicitation documents, which is usually about 30 minutes after the submission time. To prepare for the bid opening, the bids are collected and taken to a designated venue where the opening event will take place. Bidder's, and the public, should be allowed to witness the opening. The opening should take place at the appointed time, and delays should be avoided. This gives anyone attending the bid opening event the indication that it is taken seriously by the procuring entity and lends credibility to the process.

The purpose of the bid opening event is to open the bids received, read out the names of the bidders and record the bid price for all participants to see. This may be done physically or electronically but should be done as was indicated in the solicitation documents and not be arbitrarily decided by the coordinator of the opening event. The coordinator of the opening event will prepare a checklist of the bid opening process. The bid opening event should be carried out without any discussions with those attending the event, and no decisions (except for rejecting late bids) should be made at the bid opening event.

Before the opening session begins, attendees should fill out an attendance list. The remainder of the process is summarized as follows:

1. Bid price from the letter of bid is read out and written down (electronically or manually) for all present to see
2. A bid opening checklist is prepared with the following information:
   a. Was the bid submitted on time?
   b. Was there any "Substitution", Withdrawal", or "Modification" of any bid submitted?
   c. Is the outer envelope of the bid sealed?
   d. Is the bid form completed and signed?
   e. Were there any alternative bids?
   f. Is the bid validity date indicated?
   g. Is documentary authority (PoA, etc.) for signing the bid enclosed?

h. Were any discounts offered?
i. If required, was the bid security submitted and what is the amount?
j. Is the bidder a Joint Venture or Sole Provider?
k. Name and title of bidder's representative attending the opening?

The results of the bid opening event are then recorded in the bid opening minutes, and a copy of the minutes is given to all present, sent to them after the end of the opening event or posted to an authorized website. In any case, bidders are usually not obligated to attend the bid opening event, but the procuring entity is usually obliged to send a copy of the bid opening minutes, which must include the bid opening checklist, if one was prepared, but more importantly, the bid prices read out during the bid opening event. The bid opening minutes should also be posted to the procuring entity's official website and a copy sent to the donor entity (in the case of donor-funded procurements) for their review, if the procurement method requires prior review.

37. What is the process for receiving and opening proposals for consulting services?

Receiving and opening proposals for consulting services is different from receiving bids for goods, works and non-consulting services procurement. The submissions for consulting services are called proposals and they are submitted in accordance with the instructions given in the Request for Proposals. When the proposals are opened, the financial proposal remains sealed until the results of the technical evaluation are obtained, and the technical evaluation report is signed by the evaluation panel and approved by the competent authority.

Consulting firms submit two proposals, one technical and one financial. These proposals may be submitted at the same time or at different stages depending on the procurement method. If submitted simultaneously, they should be sealed in two separate envelopes. The purpose of the opening is to confirm that the proposals are in two separately sealed envelopes. Only the technical proposal is opened, and the technical proposal submission form checked to determine if it is signed by the authorized representative of the firm. The financial proposal is checked to determine if it is unopened, and it is separated and secured until the financial proposal opening of all the companies that scored at least the minimum qualifying mark on their technical proposal.

38. What is a bid security and when is it used?

A bid security is a monetary guarantee that is calculated as a percentage of the budget estimate of a procurement requirement or a percentage of a bidder's bid price. It is used by the client/ purchaser as a protection against bidders withdrawing their bid prior to the end of their bid validity period, or for refusing to sign the contract if recommended for award.

During the bidding process, for goods, works and non-consulting services, it is customary to require bidders to submit a bid security or bid securing declaration along with their bid. This may be necessary for the procuring entity to be reasonably assured that bidders would be discouraged from withdrawing their bid or refusing to sign the contract, if they are accepted for contract award.

A bid security may be a fixed monetary amount or a percentage of the bid price. It is usually less than 5%. The format and amount should be stipulated in the governing procurement rules and clearly stated in the solicitation documents.

Some of the acceptable formats of the bid security are:
1. unconditional bank guarantee,
2. certified check, or
3. bond.

39. What is a bid securing declaration and when and why is it used?

The bid securing declaration is a non-monetary form of bid security. It is a notarized statement made by a bidder committing to sign the contract if they are selected before the end of the bid validity period indicated in the solicitation documents. In this notarized statement, the bidder agrees to be disqualified from bidding for any future government contracts, for a stipulated period, if they either withdraw their bid, fail to sign the contract before the end of the bid validity period, or are unable to submit a performance guarantee, if required. The bid securing declaration usually only apply to the procurement of goods, works and non-consulting services.

40. What is the purpose of the bid security and the bid securing declaration?

The main purpose of the bid security and the bid securing declaration is to discourage bidders from withdrawing their bids before the end of the bid validity period or from refusing to sign the contract if they are recommended for award. Any of these actions by the bidder could result in delaying the procurement process, which could delay the delivery of public goods and services and result in wasting public funds due to the time and effort invested in executing the procurement process.

41. What are the three differences between the bid security and the bid securing declaration?

1. A bid security requires a bank guarantee, while the bid securing declaration requires only a notarized statement.
2. The bid security implies a possible material loss in case it is forfeited, while a bid securing declaration entails a potential loss of future bidding opportunities.
3. The bid security may result in a direct monetary loss to the bidder, while the bid securing declaration may result in an opportunity cost.

42. What are the advantages and disadvantages of using the bid security versus the bid securing declaration?

Advantages

The bid security reasonably assures the procuring entity that the bidder is serious. If bidders are eligible, qualified, and their prices are reasonable, there is a greater likelihood that a contract will be signed.

A bid securing declaration allows qualified bidders to participate in the procurement process without incurring the cost of obtaining a bank guarantee for the bid security, and thus may increase the level of participation in the bidding process.

Disadvantages

1. Some bidders may be deterred from bidding if they consider the bid security amount is too high.
2. The cost of the bid security may be added by the bidder to their bid price to cover the expense incurred.
3. Some qualified bidders may not be able to afford the bid security, resulting in a reduction of the number of bids received, effectively restricting the competition.

· CHAPTER 7 ·

EVALUATING BIDS AND PROPOSALS

43. When does the bid or proposal evaluation start?

The evaluation process starts either after receiving quotations, or after receiving and opening bids or proposals. Quotations don't require formal opening. The same is true for individual consultants' selection. Bids and proposals do require a formal opening event. The evaluation process should be scheduled to begin immediately after the opening event.

44. How does one prepare for bid or proposal evaluation?

To prepare for the evaluation of bids and proposals an evaluation panel needs to be nominated and approved by the competent authority. Evaluation panel members should be technically qualified in the subject matter of the procurement and be available and committed to concluding the process in a timely manner to avoid undue delay in contract award. The lack of availability of evaluation panel members is a major cause of delay in the procurement process. So, having members that are dedicated to the evaluation process is important to getting it done quickly.

Also important are evaluation procedures for the evaluation panel to follow. These procedures may have sample scoring sheets for the panel to use in developing those appropriate for the evaluation being undertaken. If there is a coordinator assigned to oversee the process and advise the panel, the coordinator would prepare the template scoring sheets and review the procedures to ensure they are relevant, if not, the procedures may be updated based on the evaluation criteria, and the coordinator would brief the evaluation panel on the procedures and use of the scoring sheets before beginning the evaluation.

In addition to the procedures and scoring sheets, in some cases a non-disclosure agreement is prepared and signed by each evaluation panel member. Its purpose is to identify perceived or actual conflict of interest.

45. What is the basis for the evaluation of bids and proposals?

Evaluation criteria are used to evaluate bids and proposals. These criteria are determined before inviting bids and proposals and are included in the solicitation documents. Only the evaluation criteria stated in the solicitation documents may be used for evaluation purposes. To use any other criteria would go against the principles of transparency and integrity. Bidders must be aware of how they will be evaluated and what is the basis for the evaluation of their bid or proposal. Knowing this helps them determine if they qualify for the procurement and know what qualifications are required to increase their chances for contract award. Additionally, to evaluate bids or proposals using criteria other than those stipulated in the solicitation documents could lead to a bid challenge, dispute or cancellation of the process for failure to follow the procurement rules and fundamental principles.

46. When is the evaluation panel selected?

The need for an evaluation panel and their composition is something that should be considered already during the procurement planning stage, and more so when scheduling procurements, because delays in the evaluation process often result in delays in awarding the contract. There are frequent delays in this process, either because the evaluation panel does not have the technical expertise for the procurement they are evaluating, they have other commitments and cannot fully commit to the evaluation process, or there is delay in contacting them to initiate the evaluation process immediately after concluding the bid opening event.

When selecting the evaluation panel, the panel members' experience with similar procurements needs to be considered, also if they are fully available throughout the evaluation process. A good procurement schedule helps to alert panel members of when they need to be available and what the procurement is about, by giving them a copy of the solicitation documents or request for proposals that would be the basis for the procurement they will be evaluating.

47. What is the composition of the evaluation panel?

The evaluation panel (also called evaluation committee) should be composed of at least three members. The uneven number of panel members is important especially in the event of disagreement between panel members. At least one member should have technical expertise in the procurement in question, another should preferably have some knowledge of procurement and the evaluation process, and what are the obligations, responsibilities and limitations of the panel members. The third member could be from the requesting entity and should ideally have knowledge of the evaluation process and have participated in a previous bid evaluation.

Panel members should preferably be of similar hierarchy. Supervisors and subordinates should not be on the same panel. This avoids any undue pressure on subordinates to comply with the wishes or share the opinion of their superior. Panel members should be independent and free to dissent with the opinion of other panel members.

48. Is it good practice to nominate a chairperson of the evaluation panel?

The evaluation panel should be composed of individuals that have the same level of authority on the evaluation panel. No member should be subordinate to another, as this would undermine the fact that each member has the right to evaluate without any tacit or actual pressure from another member. Each member is to evaluate based on their understanding of the evaluation criteria and to what extent the bidders bid, or proposal meets those criteria. As a result, it is not good practice for one of the panel members to be named chairperson because this gives the impression that the chairperson is the one who decides the outcome of the evaluation process, when the decision should be by consensus of the panel members, or by averaging the scores, or by majority vote, whatever is allowed under the procurement rules. If under your procurement rules a chairperson is allowed, care must be taken to ensure they have only a coordinating role and speak only on behalf of the evaluation panel and the consensus reached by the panel. They should not speak only what is their opinion or leave the other members no other choice but to follow what they have decided.

49. What are the requirements for nominating evaluation panel members?

Evaluation panel members need to be identified and selected before the submission of bids or proposals, so that once the bids are opened they can immediately begin the evaluation process to avoid delaying the completion of the procurement process.

Evaluation panel members need to be technically qualified in the subject matter of the procurement requirement. Ideally some, or at least one, should have participated in the preparation of the technical specifications or terms of reference, so they have a good understanding of the requirement and what is expected of the contractor, supplier or service provider.

Evaluation panel members must commit to being impartial during the evaluation process and to keep all information confidential, revealing the results only to authorized persons with a need to know. Their decision should not be affected by another panel member, because the scores arrived at, should not be based on anything else besides the facts presented to them in the bids or proposals received.

50. Who selects the evaluation panel?

The selection of the evaluation panel members should be a joint effort between the procuring entity and the requesting entity; but the procuring entity should decide, unless otherwise stipulated in the procurement rules. Usually, the requesting entity would nominate members for the procuring entity's review and approval. The curriculum vitae or summary of qualifications of nominated evaluation panel members is sent to the procuring entity (or designated authority) for review and approval. On donor-funded procurements, the evaluation panel may need the approval of the donor entity as well.

51. What is a confidentiality agreement?

A confidentiality agreement, also called declaration of impartiality and confidentiality, is a requirement of some donor agencies and governments. It is a document signed by all members, observers and coordinators of evaluation panels. Its purpose is to identify and preclude any potential or actual conflict of interest. Anyone involved in the evaluation process must declare any affiliation, business, family or amicable, with any bidder that has submitted a bid or proposal for the procurement requirement under evaluation. The declaration of impartiality and confidentiality is a legal document. False declarations could result in legal or disciplinary action.

Any panel member that has a potential or actual conflict of interest is encouraged to declare it, and a decision made by the competent authority whether the panel member can continue or should be removed from the evaluation panel. There is nothing inherently wrong with a panel member having a conflict of interest, provided that it is declared by the individual and resolved by the competent authority.

52. How long does it take to evaluate bids and proposals?

It is difficult to determine how long the evaluation process should take because it depends on the complexity of the procurement requirement, the technical expertise of the evaluation panel, how complete the bids or proposals are (which reduces the need for requesting clarification), if the evaluation panel convenes daily, and the number of bids or proposals received. A rough estimate is one and half to two days to complete the evaluation of one bid or technical proposal. Bids and quotations received for small value and less complex procurements take less time.

The procurement method also affects the duration of the evaluation process, because some methods require the completion and approval of a technical evaluation report as a prerequisite to doing the financial evaluation. Another factor influencing the timeframe for completing the evaluation process is the period it takes the approving authority to approve the evaluation report once it is submitted to them.

53. What is a bid or proposal validity and what impact does it have on the selection process?

The bid/proposal validity period is the number of days a bidder agrees to keep their bid/proposal valid. The solicitation documents must state the required bid validity period, and bidders must confirm that their bid/proposal is valid for that period. Bids/proposals with less validity are rejected as non-responsive to the stipulations of the solicitation documents.

The validity period is usually set by the procuring entity in such a manner as to allow sufficient time for completion of the evaluation process, approvals and contract signing.

Whenever it is not possible to sign the contract before the validity period expires, bidders should be formally requested to extend the validity period. If they agree, the process should continue. If they don't agree, their bid/proposal should not be further considered, and their bid security should be returned after the selection process is concluded. Bidders are not obliged to extend the validity of their bid if the evaluation and selection period extend beyond the bid/proposal validity date.

54. What is a bid or proposal evaluation report?

Bids and proposals are evaluated after the bid/proposal opening event. The results of the evaluation are recorded in an evaluation report, which includes summary details of the selection process, how the evaluation process was carried out and the award recommendation.

The evaluation process ideally results in a recommendation for contract award to a qualified and responsive bidder that offered the required solution consistent with the solicitation documents.

Preparation of the evaluation report is the responsibility of the evaluation panel. The report is prepared after the evaluation panel concludes the scoring, depending on the procurement method used, and after compliance with the technical specifications, eligibility requirements, completeness of the bids, and responsiveness to the solicitation documents has been determined. Upon concluding the evaluation process, the evaluation panel signs the evaluation report and it is sent for approval before proceeding with contract negotiations or award, if required.

55. Who approves the bid or proposal evaluation reports?

After the evaluation report is completed and signed by the evaluation panel, it is sent for approval. The approving authority may be a tender board, contracts committee or a donor entity. With respect to government procurement, there may be a need for several approvals depending on the complexity and monetary value of the procurement. If the procurement approval is at a ministerial level, a departmental tender board may need to approve before the ministerial tender board. Likewise, If the procurement approval is at the cabinet level, there may be a need for prior approvals by both departmental and ministerial tender boards. This depends on the governing procurement rules.

56. What is the process for preparing the contract?

After concluding the evaluation report and obtaining the required approvals, the contract should be prepared. The form of contract contained in the solicitation documents is the one that is used. Ideally, the drafting of the contract may begin before finalizing the evaluation report, but at that stage the draft contract cannot be finalized because the details of the winning bidder are unknown. The procuring entity is responsible for drafting the contract, using the form of contract found in the solicitation documents, but the final draft is done once the evaluation report is finalized and approved. If there is no form of contract in the solicitation documents, then the form of contract stipulated in the procurement rules shall be used.

In the case of goods, works and non-consultant services procurements, a draft letter of acceptance is also prepared along with the draft contract, and once approved, both the approved letter of acceptance and the contract are sent to the winning bidder informing the results of the evaluation process and giving them a reasonable time to sign and return the contract to the procuring entity.

With respect to consulting services, after the evaluation report is approved, the consulting firm is invited to contract negotiations, during which the draft contract is reviewed and agreed between the chief negotiators of the parties. After the negotiations, both chief negotiators will put their initials on the draft negotiated contract, which is then sent for approval before signing.

Before issuing the letter of acceptance or signing the negotiated contract, a notification of intent to award may need to be sent to all bidders that submitted bids/proposals. Other requirements of the procurement rules need to be followed. This may include publication of the intent to award in newspapers and posting to official websites. In all cases, what is stated in the solicitation documents should be followed.

· CHAPTER 8 ·

CONTRACT NEGOTIATIONS AND AWARD

57. What are contract negotiations?

Contract negotiations is often misunderstood as price bargaining. In fact, in public procurement, price negotiations are limited to circumstances where the price is not a determinant of contract award. Where the evaluation of the price, after the technical evaluation, is what determines the recommended bidder, price negotiation is not permitted, because changing the price, could affect the outcome of the evaluation process. This is applicable in procurements where the lowest price, after technical evaluation, determines the recommended bidder (as with least cost selection methods), also in cases where the sum of the combined technical and financial scores determines the recommended consultant (as with the quality and cost selection method).

However, in cases where technical qualifications are what determines award recommendation, the proposed price may be negotiated if considered unreasonably high, based on factual evidence or historical data (past contracts with the consultant or firm for similar requirements, for example). The hiring of individual consultants, where the highest technically qualified consultant is invited for contract negotiations, is an example of a case where the price may be negotiated if found unreasonably high. This is because the selection is primarily based on the consultant's qualifications and not the price. The same is true for single source procurements; because there is no competition, it is important to ensure the contract price is reasonable.

Although, negotiations are not always permissible or required, and can include price negotiations only under limited circumstances, negotiations are usually permitted for consulting services procurements and not the procurement of goods or works, unless otherwise indicated in the procurement rules or solicitation documents.

58. When does contract negotiations take place and what does it involve?

As mentioned before, negotiations are mostly permitted, and required for consulting services procurements, and are applied in the hiring of consulting firms and individuals. Negotiations are initiated after the financial evaluation or a combined evaluation report is completed and approved. The recommended consultant is then invited for contract negotiations, which can be either face to face, via email or teleconference.

The procuring entity prepares the negotiation points, which include weaknesses observed in the personnel, methodology, work plan and other aspects of the technical proposal of the recommended firm. On the other hand, the consultant may want to discuss issues related to the terms and conditions of the contract, insurance, improvement to the terms of reference, mobilization of resources, personnel availability, payments processing, etc.

Each side, the client and the consultant, must name a negotiating team, and the chief negotiator of each must have authority to reach a final decision on the draft negotiated contract.

At the end of the negotiation process, minutes of negotiations are prepared and signed, and they are annexed to the draft negotiated contract (if permitted under the Procurement Rules), which is initialed by both chief negotiators before concluding the negotiations process. The draft negotiated contract is then sent for approval, and the notification of intent to award is published after the formal approval is granted; then, the contract is signed by the authorized representatives of both parties.

59. What is the contract award process for goods, services, and works?

Once the evaluation report, with award recommendation, is approved by the competent authority, the process of contract award for goods, works and non-consulting services, and consulting services is slightly different.

Goods, Works and Non-Consultant Services

To begin with, contracts for goods, works and non-Consultant services are usually not negotiated. These procurements follow the same process: after concluding bid evaluation, a draft contract and letter of acceptance is prepared, and once approved, they are sent to the winning bidder, allowing them a set number of days to sign and return the contract.

Before that, a notification of intent to award is prepared and sent to all bidders that submitted bids, and it is also published in authorized venues and posted to official websites. If there is a bid challenge system in place, there may be a requirement to delay the issuance of the acceptance letter for a set number of days (stand-still period) to allow for any protest. Once the stand-still period has passed without any protest/bid challenge, the letter of acceptance may be issued. The letter of acceptance is a notification of award, which is submitted to the winning bidder with an unsigned contract. It allows the bidder several days, usually 28 days, to finalize stipulated details (submit performance guarantee, warranty, etc.), sign and return the contract.

Consulting Services

For consulting services (hiring firms and individual consultants), once negotiations are successfully concluded and the formal approval of the negotiated contract obtained, an intent to award notice is sent to all firms or individuals that participated in the process. If there is a requirement for a stand-still period to allow for bid challenges, such period must elapse before the contract is signed by the parties.

60. What is the bid challenge process and what impact does it have on the evaluation and selection process?

A bid challenge occurs when a bidder submits a claim to the procuring entity or competent authority, alleging that they were adversely affected by an action or inaction of the procuring entity during the selection process of a specific procurement requirement. Bidders have the right to seek review of the procurement process, and a remedy, if they believe the procuring entity failed to comply with the procurement rules throughout the procurement process for the procurement in which the bidder participated. Upon receipt of the bid challenge, the procurement process may be suspended until the bid challenge is resolved.

Bid challenge procedures are formal. They indicate how the challenge should be submitted, when and where. There are two review levels. The first-level review is done by the procuring entity (or another if stipulated in the Procurement Rules) and must be completed within a stipulated number of days after receiving the bid challenge. This first-level review may be appealed and will then be sent to an independent review panel, which has the authority to override the first-level decision. The decision of the independent review panel is final.

61. What is the debriefing process and what are the different ways in which it can be carried out?

A debrief is used to inform bidders of the details of the evaluation process with respect to the strengths and weaknesses of their bid or proposal. It may be done orally or in writing at the bidder's request, unless otherwise stipulated in the procurement rules.

The information in the debrief report should be enough to inform the bidder of the strengths and weaknesses of their bid, proposal or application, and what led to their not being recommended for contract award.

The debriefing report should be prepared within the time-frame stipulated in the procurement rules and should address only the details of the requesting bidder.

· CHAPTER 9 ·

CONTRACT ADMINISTRATION BASICS

62. What is contract administration?

Contract administration begins as soon as the contract is awarded and focuses primarily on the management of risks in contract implementation. Its main purpose is to monitor contractor performance to ensure the objectives of the contract are met on time and within budget, and to detect potential problem areas and resolve them before it is too late. Payment for satisfactory compliance with contract terms and conditions is also part of contract administration, as is the management of contract variations, amendments and the suspension, termination, conclusion and formal closing of the contract.

63. What are some of the key elements of contract administration?

The following are 14 key elements of contract administration:

1. Contract administration planning
2. Contract commencement
3. Kickoff meetings
4. Mobilization of resources
5. Record keeping
6. Performance monitoring
7. Payments Processing
8. Environmental, health and safety considerations
9. Contract variations and amendments

10. Contract suspension
11. Dispute resolution
12. Contract termination
13. Contract completion
14. Contract closeout

64. Is there a difference between procurement and contract administration?

The public procurement process begins with assessing needs and determining requirements for supporting government operations and providing public goods and services. The procurement process ends with contract award. Delivery of goods and services are not part of the procurement process; they are more related to logistics and supply chain management.

Contract administration begins from the moment a contract is signed and the need for monitoring the implementation of the contract becomes a reality. The administration of the contract covers many aspects; primarily: receiving and acceptance of goods and services, performance monitoring, payments, record-keeping, contract modification, termination, completion and close-out, to mention a few.

65. Who is responsible for contract administration?

The primary responsibility for contract administration lies with the accountable official that signs the contract. This is often delegated to another official for operational purposes. However, to be effective, contract administration should be a joint effort; not only the responsibility of one entity. The extent of their participation should be in proportion to the value each entity adds to the process, given the importance of their expertise to the effective administration of the contract. So, it is important to develop a comprehensive contract administration plan with the participation and agreement of all entities involved, and with a thorough understanding of the responsibilities each has in the process.

66. What is contract administration planning?

The contract administration plan identifies and records all deliverables due over the life of the contract, and how inspections and performance monitoring and reporting will be carried out. Contract administrators must be familiar with all contract milestones, including its terms and conditions, to properly monitor and assess progress, and identify any problems as they arise. The development of the contract administration plan and the nomination of the contract administration team should be done during the selection process, before contract award. The nominated contract administrators should participate in the development of the contract administration plan. They should also be given a copy of the contract as soon as it is signed.

67. What is contract commencement and how does it differ from the effective date of the contract?

Contract commencement and contract signing date does not always coincide. Contract signing usually happens before contract commencement. Contract signing is the moment at which the contract becomes binding between the parties. It also signals the beginning of the contract administration phase. The simultaneous occurrence of contract signing, and contract commencement depends on the procurement category (goods, works, or services) and the complexity of the requirement.

The procurement of readily available goods is relatively straightforward, and the signing and commencement of the contract is usually simultaneous, unless there is agreement to delay commencement until the vendor receives an advance payment and the purchaser a corresponding advance payment guarantee or a performance guarantee. For works and services procurements, in contrast, it is common practice to allow contractors and service providers up to 30 days after the signing of the contract to mobilize personnel and equipment to the location where the contract will be implemented.

68. What are kickoff meetings and why are they important?

Kickoff meetings are usually scheduled for more complex procurement requirements. Their purpose is for the purchaser/employer to familiarize the contractor, supplier or service provider with the details of the contract administration process, and to introduce the person or entity responsible for managing contract implementation. The contractor is expected to present a preliminary mobilization plan and a conceptual contract implementation plan. Kickoff meetings are usually held after all the formalities of advance payments and performance guarantees have been resolved, and before beginning the works or service delivery.

69. What is mobilization of resources and what is its purpose?

Mobilization is the period between contract signing and contract commencement agreed between the purchaser and the supplier, contractor or service provider. It is the period after contract signing that the contractor dedicates to getting all personnel and equipment in place, including facilities taken over to commence works or services provision. The mobilization period ends when the works or services begin.

During mobilization of resources, the service provider moves equipment and personnel to the project site or begins service delivery from their home office. Preparatory work on the contract is initiated and clarifications on the format and content of the inception report, and other concerns are sorted out during this period.

Additionally, during this period matters concerning performance guarantees, advance payments and corresponding securities are sorted out. Performance guarantees are unusual for consulting services, but can be applied depending on the complexity, risk and monetary value of the contract.

Mobilization is hardly necessary for the provision of goods, but if services are included, such as for requirements needing supply, installation and commissioning, a mobilization period may be necessary.

70. What is record-keeping and why is it important?

In procurement, record-keeping has to do with the development and maintenance of procurement and contract files to ensure and verify compliance with the procurement rules and contract terms and conditions.

The keeping of procurement and contract records may be electronic or physical, sometimes both. They include all documents relevant to the pre-tendering, tendering and contract administration phases. Every event in the procurement process must be recorded and all records filed in such a manner that, if necessary, the entire procurement and contract administration processes may be reconstructed from these records. Procurement and contract records are also important for maintaining an audit trail of each procurement requirement, from the initial receipt of the procurement requisition up to the closing of the contract.

· CHAPTER 10 ·

CONTRACT IMPLEMENTATION INSTRUMENTS

71. What are different types of guarantees and why are they important?

Guarantees are used in public procurement to protect the purchaser (buyer) against non-performance or substandard performance of a contract (performance security) or against any action that may put the selection process at risk (bid security). The most common form of guarantee is the bank guarantee.

There are various types of guarantees: performance guarantee, unconditional bank guarantee, demand guarantee, performance bond, bid bond, and irrevocable letter of credit. Regardless of the type or format of the guarantee, they all give the purchaser the right to demand payment of the guaranteed amount, under expressly stipulated circumstances, if the other party takes or fails to take an action that was clearly indicated in the guarantee.

Guarantees are usually in the form of a monetary payment to the purchaser for the contractor's breach of an obligation. Additionally, the performance guarantee can be in the form of retention of an agreed amount from periodic payments to a supplier or contractor over the life of the contract.

Insurance is another form of guarantee. It can be used to secure goods and services against loss, damage or other forms of risks associated with the implementation of the contract. With respect to consulting services, firms and individuals are expected to have indemnity insurance coverage as a guarantee against personal and third-party liabilities, and any damage suffered by the purchaser resulting from the services rendered by the consultant.

72. What is a performance security?

A performance security is usually required for infrastructure works contracts, and to a lesser extent for services or goods contracts, except for complex requirements of high monetary value. Performance securities and their condition of use should be clearly stipulated in the procurement rules and in the solicitation documents.

The purpose of the performance security is to ensure the supplier, contractor or service provider perform according to the stipulations of the contract. If their performance is deficient, the guarantor of the security is obliged to pay the employer/purchaser the guaranteed amount of the security, which is usually ten percent of the contract value.

A performance guarantee may be issued by a bank, insurance company or other institution acceptable to the purchaser, it should also remain valid for the full amount until contract completion.

In the case of works procurement, when the work is completed, the performance guarantee may either be converted (with the agreement of the contractor or as may be stipulated in the contract) to a defects liability guarantee. The defects liability period (usually one year) begins upon completion and acceptance of the work performed under the contract. It is used as a guarantee for any defects that may arise in the infrastructure within this period, and which are attributed to the standards of quality of the work completed by the contractor.

73. What is a letter of credit?

A letter of credit is a promise to pay. It is purchased by the purchaser from a bank. It is an agreement to pay the supplier on behalf of the buyer/purchaser under the conditions stipulated in the letter of credit. Once the supplier complies with the conditions and presents the agreed documentation, the bank is obliged to pay them as indicated in the letter of credit.

The letter of credit is used primarily for goods procurement. It is negotiated to ensure suppliers get paid in a timely manner after performing according to the contract or purchase order.

If, for example, the buyer/purchaser agrees with the supplier to pay once the goods are placed on-board a ship, the supplier only needs to present the agreed documentation to the bank as proof that the goods were loaded on the ship, and the bank is obligated to pay the supplier the amount stipulated in the letter of credit.

74. What is a warranty and how many types are they?

A warranty is also a form of guarantee. It is a written assurance, by a contractor, supplier or service provider, of the integrity of the contract deliverable and their responsibility to repair or replace a good or re-do the work free of charge to the purchaser. The conditions and duration of the warranty must be stipulated in the contract

With respect to goods procurement, a warranty is a written assurance that if goods malfunction or are defective within a stipulated period (if the user is not at fault), they will be repaired or replaced, free of charge, by the supplier or manufacturer. Sometimes, an agreed amount of the contract value is retained by the purchaser until the warranty period expires. This must be stated in the contract.

In the case of works procurement, the defects liability period is a form of warranty. The defects liability period is protected either by the retention of the contractor's performance security and corresponding bank guarantee, at the end of the contract performance period. The defects liability period can also be covered by retentions made from payments to the contractor over the life of the contract.

The defects liability security is kept for a period, usually one year, for the repair of any defects found in the infrastructure built or refurbished by the contractor. If a defect in the infrastructure is detected during the defects liability period, the contractor is notified and given the opportunity to repair the defect. If the contractor is unable or unwilling to repair such defect, the purchaser has the right to use the funds retained from the contractor for such purposes.

Once the defects liability period expires, if no defects were discovered or those identified were satisfactorily repaired by the contractor, the retention money is returned to the contractor, as stipulated in the contract.

75. What are International Commercial Terms and what is their purpose?

International Commercial Terms (Incoterms) are agreements made for the purchase and shipment of goods, primarily from international suppliers. They are a series of predefined commercial terms that are published by the International Chamber of Commerce. Incoterms define the responsibilities of buyers and sellers for the delivery of goods, and they determine how costs and risks are allocated.

The purpose of the Incoterms is to reduce confusion over the interpretation of shipping terms, because they outline exactly which party is obligated to take control of and insure the goods at different stages in the shipping process.

Incoterms also specify which of the parties has the obligation for clearing the goods for export or import, and for packing the goods according to requirements. Further details may be found at the Incoterms website.

76. What Purpose does insurance serve during contract implementation?

The need for insurance is a requirement under contracts for all procurement categories. The insurance requirement is for protection against events that could affect or impede contract performance.

Goods procured should be fully insured against loss or damage incurred throughout the manufacturing, shipment and delivery processes. The supplier needs to have insurance against any risk of liability related to the performance of the contract. The requirement for insurance coverage is also specified in the Incoterms, and in the solicitation documents.

Civil works procurement contractors are required to provide insurance covering the period from the start date of the contract up to the end of the Defects Liability Period (sometimes beyond), to cover risks of personal injury, and loss of or damage to works, plant, materials, equipment, and property in relation to the contract.

Non-consulting services insurance is similar to what is required for goods procurement.

Insurance is also required under consulting services contracts undertaken by companies as well as individuals.

· CHAPTER 11 ·

CONTRACT MONITORING

77. What are environmental, health and safety considerations, and Why are they Important?

Environmental, health and safety considerations are provisions stipulated in the contract to safeguard the environment, and the health and safety of the employees of the contractor and third parties that may be affected by the contractor during the implementation of a contract.

Contractors, suppliers and service providers are expected to take precautions to protect the health of their personnel and to maintain the safety of the work environment for their personnel and third parties as well. There are contract clauses which address environmental, health, and safety issues, and contractors, suppliers and service providers are expected to comply with them. They are also required to follow and implement accepted procedures in compliance with environmental, health, and safety standards. Such requirements are stated in their contracts, sometimes with direct reference to a specific government text or norm, or that of a donor entity or a national/international environmental, health, and safety requirement.

78. What is the importance of inspections and testing in the receiving process?

Inspections are important to determine if goods received, works completed, or non-consulting services delivered, are in accordance with the technical specifications and terms and conditions of the contract. Testing takes place based on conditions specifically stipulated in the contract.

In goods procurement, inspection and testing may be done during the manufacturing process. This also must be stipulated in the solicitation documents and the contract. Inspection of goods during the receiving process is essential to confirm that the goods are received in the conditions and according to the specifications indicated in the contract. Testing during the receiving process is rare, and because it may result in the destruction of the sample, testing is usually done at the manufacturer's facilities.

In works procurement, inspection is frequently done at certain intervals to ensure that the works are being carried out in compliance with the contract. Testing is also done when required by the contract. In civil works, where concrete is used, for instance, there is usually a requirement for the supervising engineer to do a compaction test of a sample of the concrete to determine if it complies with the contract specifications. The gauge of the steel used in the building of a structure may also be tested to verify compliance with the contract. In some cases, only inspection is done because the work carried out does not lend itself to testing, such as in the case of minor repairs and maintenance.

Only certain services, mainly non-consulting services, allow for inspection and testing. Certain maintenance and repair services, for example, may be inspected and tested; such as air-conditioning and motor vehicle maintenance and repair, to determine their condition before and after servicing. Consulting services, in contrast, are very difficult to inspect and test to determine the quality of the services provided, since the result of the service delivery is primarily intangible.

Inspection and testing must be planned. The parties involved must be aware of their role. The place for inspection and testing must be determined (if not already stipulated in the contract) and communicated to all involved in the process. When, before shipping goods, inspection and testing is done at the manufacturer's facility, a purchaser's representative may be present to witness the event. The purchaser's representative should be qualified and able to understand the actual inspection and testing process; the purchaser may hire a specialized firm to undertake this function.

The receiving and inspection that takes place at the destination, before acceptance, as stipulated in the contract, must also be done by a qualified person able to verify if the goods being received meet the technical specifications and are in a condition that is acceptable for their intended purpose.

The procuring entity may participate in the receiving and inspection process to take note of what was received, when, where and in what condition. The primary responsibility for receiving and inspection should be a designated entity or agent (hired by the purchaser) who has the technical know-how to confirm the technical configuration and condition of the goods or services being received.

79. What is performance monitoring and evaluation and how is it done?

Performance monitoring and evaluation are all actions taken by the contract administrator to ensure the contract is implemented in accordance with the terms and conditions. Day-to-day contract administration is carried out by an individual or entity (line manager, project manager, or implementing entity) in coordination with the procuring entity. The contract administrator is responsible for the oversight of all contractors, suppliers and service providers, and monitors their compliance with the terms and conditions of the respective contract. The contract administrator is also responsible for promptly informing the procuring entity of any delays in delivery or other forms of non-compliance with contract terms and conditions, with a recommended action to be taken. The procuring entity, in coordination with the contract administrator, prepares any draft variations or amendments to the contract, which after review and approval are signed between the respective parties to the contract.

80. What is the purpose of monitoring goods procurement contracts and how is it done?

Performance monitoring of goods procurement contracts is primarily reduced to receiving and inspection (sometimes testing), ensuring the right goods are received on time and fit for their intended purpose, and that the warranty and warranty period are as indicated in the contract or purchase order.

After contract commencement, the supplier is responsible for fulfilling all the conditions stipulated in the contract; including, meeting the delivery, warranty and insurance requirements, and providing the goods according to the specifications stipulated in the contract. Whenever performance testing of the goods is required, the contract must indicate when and where the performance testing will take place -before the shipment or before the receiving (or commissioning) process.

For complex goods, such as machinery, generators, vehicles, and so forth, a technical specialist or inspection agent should be engaged to perform the receiving and acceptance function to ensure the goods conform to the specifications of the contract.

During the receiving process, in the event of any deficiency in the volume, quality or compliance with the technical specifications provided in the contract, a discrepancy report must be prepared by the receiving agent or entity for submission to the supplier. The nature and cause of the discrepancy must be determined and indicated to justify the claim of the receiving entity. A copy of the discrepancy report should be placed in the contract files.

After the goods have been satisfactorily received, the warranty period begins. It is important for the contract administrator to be aware of the conditions of the warranty and to monitor the condition of the goods to identify any defects that may justify the need to apply the warranty. In such cases, immediate action must be taken to seek remedy from the supplier; which could result in the repair or replacement of the goods.

If final payment to the supplier was retained as a warranty security, the expiration date of the warranty period must be monitored to ensure the release of the final payment to the supplier once the warranty period is completed.

81. What is the purpose of monitoring works procurement contracts and how is it done?

Works procurement usually requires design, construction and specialized supervision. In public procurement, civil works is mostly contracted out; while the design and supervision may be done by a Government entity such as Public Works. Alternatively, the entire requirement is contracted out, as is usually the case in project procurement. The arrangement, depending on the complexity of the requirement, could be that the design and the supervision of the works are done under one contract, and the construction under another. Another option is to do the design and construction of the works under one contract (design & build), and the construction supervision under a different contract. A final arrangement is for all three requirements: design, construction, and supervision to be carried out under separate contracts, each with different firms.

Regardless of how the civil works requirement is fulfilled, the supervising engineer has the contract management responsibility over the construction contract (or design and build contract), on behalf of the purchaser (employer); and they must ensure the works are carried out in accordance with the technical specifications of the contract.

The implementing entity monitoring the supervising engineer's contract should alert the procuring entity of any breach or non-compliance with the contract terms and conditions so appropriate corrective and timely action are taken when needed.

While it is the responsibility of the supervising engineer to monitor the works contractor's performance, the monitoring of the supervising engineer's performance is the responsibility of the implementing or requesting entity.

82. What is the purpose of monitoring services procurement contracts and how is it undertaken?

Services are commonly divided into two categories: consulting services and non-consulting services. This classification is primarily based on the output of the service provided. The output of consulting services contracts is primarily intellectual or mostly intangible in nature, while the outcomes of non-consulting services are mostly tangible.

With non-consulting services, a warranty on the services should be agreed, and the results (outputs) inspected and tested; however, the effectiveness of the deliverables for consulting services is sometimes difficult to determine at the completion of the contract. For this reason, consulting services providers are commonly required to have indemnity insurance that covers the firm during the assignment and for a certain period after contract completion.

Individual consultants are required to provide professional liability insurance, the details of which are specified in the contract agreed with the consultant.

In both types of services contracts, the implementing or requesting entity is responsible for designating personnel to oversee the performance of the service provider, through the monitoring of the timely submission of deliverables or completion of the services, and to identify any changes needed in the scope of the required services arising or discovered during the implementation of the contract.

· CHAPTER 12 ·

PAYMENTS PROCESSING

83. What are the different types of payments?

There are two main types of payments: advance payment and periodic payments.

Advance Payment

Advance payment is a payment that may be requested by suppliers, contractors and service providers before actual work on the contract begins. The advance payment is usually about 10 to 15 percent of the contract value but can vary depending on the agreement between the parties or limitations stipulated in the procurement rules. The advance payment is a loan to the winning bidder.

Upon negotiation and agreement on the advance payment, the supplier, contractor or service provider is obliged to present an advance payment guarantee, which may be a bank guarantee or a bond that is acceptable to the purchaser. The purchaser recovers the advance payment (repayment of the loan) by deducting an agreed amount from subsequent periodic payments to the supplier, contractor or service provider.

Periodic Payments

The periodic payments process begins after a valid payment request is received from a supplier, contractor or service provider, with a certification of acceptance of goods received, progress payment upon percentage completion of civil works, or services provided.

The payments process must be clearly stipulated in the contract, and payments promptly made, because late payments may entitle the supplier, contractor and service provider to claim interest charges on the outstanding payment amounts based on the contract terms and conditions.

For works procurement contracts, for instance, several reports are used as justification for payments processing upon acceptance by the supervising engineer: periodic performance, percentage completion, and monthly reports. A copy of such reports should be sent to the procuring entity for filing.

· CHAPTER 13 ·

CONTRACT MODIFICATION, SUSPENSION AND CONCLUSION

84. What is contract modification?

A contract modification is a change in the scope of work or terms and conditions of the contract. There are primarily two types of contract modification: (1) a contract variation and (2) a contract amendment.

1. A **contract variation** is more specific to infrastructure works contract and is used primarily when changes in the works are needed and is designed to avoid delays by allowing the site engineer to approve changes in the works proposed by the contractor whenever reasonable and contemplated in the contract. It differs from a contract amendment because it mostly applies to works contracts and is designed to prevent doing a formal amendment for all changes to the contract. In works, variations are required due to unforeseen situations, and a formal contract amendment may delay the works. So, the Contractor or the engineer can propose a variation to the works, and this may be approved by the supervising engineer. Once agreed, the works can proceed. An example of a need for contract variation could be if the design of the foundation of a structure calls for a defined depth, and during excavation it is determined that due to soil conditions the foundation needs to be deeper and piling is needed. A variation to the contract may be proposed and agreed with the site/supervising engineer to proceed with

the works, because it is something that needs immediate attention.

2. A **contract amendment**, in contrast, is a change in the terms and conditions of the contract scope of work, price, or some other contract provision. This change is usually agreed between the parties bilaterally but can also be unilateral if the terms and conditions of the contract expressly give the purchaser the right to make modifications without the supplier's, contractor's or service provider's consent.

When contract amendments are bilateral, either party to the contract can identify the need for a contract modification; but both parties are then required to agree and sign the amendment after it is approved by the competent authority.

Typical reasons for contract amendments are: contract time extension, price adjustments, changes in technical specifications and scope of the contract, and other administrative changes such as name and legal status.

The procuring entity provides guidance on contract modifications, prepares the amendment, seeks legal advice, and coordinates the process until completion. A signed amendment is kept on file, sent to the supplier, contractor or service provider, the contract administrator, and the fiscal or other relevant entities.

85. What is a contract suspension?

A contract suspension is the temporary cessation of contract implementation. It is different from the suspension of a supplier, contractor or services provider. The suspension of a contract may be invoked by either party to the contract for specific reasons defined in the terms and conditions of the contract, such as Force Majeure. The suspension of a supplier, contractor or service provider is a decision made by the purchaser which prevents the supplier from participating in the bidding process (debarment) or signing a contract. Such debarment is the result of a breach of contract or other causes indicated in the procurement rules.

Contracts should include a provision for either party to suspend or terminate the contract under certain conditions. Grounds for suspension or termination of a contract are usually indicated in the General Conditions of the Contract.

A contract suspension does not necessarily lead to contract termination. The reasons must be indicated in the contract. An example of a situation that could lead to a contract suspension is the temporary inability of a supplier, contractor or service provider to comply with the contract terms and conditions for reasons beyond their control (Force Majeure).

Whenever deficiencies are observed during contract administration, the first step is to verbally address the issues with the other party. If results are unsatisfactory, after this verbal notification, then it should be communicated formally and the other party, giving them enough time to remedy (cure) the defect. If the deficiency is not remedied within the indicated time frame, contract suspension or termination should be considered, after taking into account other available remedies.

Since one of the main objectives of procurement is to award contracts to satisfy requirements, contract suspension should be considered before deciding on contract termination. Additionally, all other possible measures should be exhausted before terminating a contract, given the cost implications, considering time lost and negative impact on service provision, of having to re-tender to select another supplier, contractor or service provider.

86. What is force majeure?

A Force Majeure event can lead to the temporary suspension or actual termination of a contract; but it cannot be considered a breach of contract because a Force Majeure event is one that is not under the control of either party. Such events cannot be prevented and are usually unforeseen. A natural disaster which causes the delay or suspension of the implementation of a contract, would be considered a force majeure event.

87. What is dispute resolution and why is it important?

A dispute is a disagreement between the parties to a contract. The primary mechanisms for resolving disputes are: amicable settlement, mediation, arbitration, and litigation.

Parties to a contract should first seek to resolve any differences through discussions to reach an **amicable settlement**. If this is not possible, mediation is the next stage.

Through **mediation**, the parties agree to the use of a third party to resolve their dispute, but because the results are not binding, the parties must agree to the results before the dispute can be resolved. If the dispute is not resolved through mediation, the next stage is arbitration.

In the **arbitration** proceedings, independent third parties are assigned to assist in resolving the dispute. The independent third parties sometimes form what is referred as a dispute board or dispute review board (DRB) or dispute adjudication board (DAB). The first task of this Board is to find a way to amicably resolve the dispute.

Litigation is where either party has the right to representation by legal counsel in a court of law.

The mechanism for resolving disputes should be clearly stated in the contract.

88. What is contract termination?

A termination clause should be included in all contracts to give either party the right to terminate the contract under certain conditions. Contract termination should be a measure of last resort.

There are three main types of contract termination: (i) contract termination for default, (ii) contract termination for convenience and (iii) contract termination as a result of Force Majeure.

Contract termination for default applies when there is a breach of contract terms and conditions by either party. Some examples are failure to deliver goods or services within the stipulated time frame, or to adequately perform construction works in accordance with the contract, also failure to make payments on time.

Termination for convenience is commonly found in Government contracts. This often-controversial clause gives the Government the right to unilaterally terminate a contract for any reason. The right to termination for convenience is increasingly being extended to the other party (contractors, suppliers and service providers) as well. Either party then have the right to terminate the contract when it is no longer in their best interest to continue with the contractual relationship. Reasons for contract termination for convenience could be because the goods or services are no longer needed, or the funding is no longer available, among other things. When termination for convenience is invoked, the party terminating the contract must adequately compensate the other, in line with the terms and conditions of the contract. For this reason, termination for convenience is risky and can be very costly for either party.

Contract termination resulting from Force Majeure may be agreed in situations where an unforeseen event, which is not under the control of either party, prevents the continuation of the contract. In such cases neither of the parties may be penalized.

89. What is contract completion?

While, contract termination may result from of a breach of the contract, a Force Majeure event or termination for the convenience of either party, contract completion signals the successful conclusion of the contract to the satisfaction of both parties, even if the contract was extended beyond the originally agreed completion date.

90. What is contract closeout and when does it take place?

Contract closing marks the end of any further contractual obligations between the parties under a present agreement. It is the point at which the contracting parties ensure all the terms, conditions and obligations under the contract have been fulfilled, and that the other contracting party has been duly compensated according to the terms and conditions of the contract.

For goods contracts, closeout formalities are completed when the goods have been delivered and accepted, and the warranty period elapsed. For services contracts, it is important to ensure that all deliverables have been satisfactorily complied with and accepted by the purchaser. Finally, contract closeout for civil works procurement is carried out only after the defects liability period has satisfactorily passed, and any performance security or retention money has been returned to the contractor.

When a contract is terminated, contract closeout should take place after ensuring that both parties have complied with their obligations under the contract, up to the agreed date of contract termination.

CLARIFICATION OF TERMS

A

Advance Payment   Payment made to a supplier, contractor or service provider before the commencement of the contract. It is requested by the selected bidder and require a bank guarantee or bond. The advance payment amount depends on the agreement between the parties and any limitations set in the public procurement rules. The advance payment is a loan made by the purchaser/employer to the contractor, supplier or services provider, and needs to be repaid. Repayment of the advance payment is done through agreed deductions from periodic progress payments made to the corresponding contractor, supplier or service provider.

Advance Payment Guarantee   A bond or bank guarantee that the selected bidder is required to give to the purchaser, as a security, allowing the purchaser to recover the advance payment in the event of the bidder's default during contract implementation.

Approval or Approving Authority
(see also: Tender Board)   The authority to approve all procurement and contract administration activities undertaken, and documents produced throughout the procurement and contract administration processes. Depending on the public procurement rules, approval authority may be delegated to a tender board, contracts committee or other institution.

**Association
(Joint Venture,
Subcontracting)**
Association is the term used to identify when two or more firms join forces to bid on a procurement requirement. There are two main types of associations: subcontracting and a joint venture. Under a subcontracting arrangement, only the principal firm is liable to the client. Conversely, under a joint venture agreement, all parties to the joint venture are jointly liable.

B

Bid

(same as Offer,
Proposal, Quotation or
Tender)
An offer from a supplier, contractor or service provider in response to an invitation for bids or request for proposals. If the offer is accepted after evaluation, the bidder recommended for award is obligated to perform in accordance with the terms and conditions of the contract.

**Bidding
Documents**

(same as Tender
Documents or
Solicitation Documents)
A set of documents issued by the procuring entity inviting offers (bids, proposals or quotations) for the selection of suppliers, contractors or service providers to fulfill specific procurement requirements.

Bid Opening
The public opening and reading out of all bids received by the designated date and time specified in the solicitation documents. The bid opening event is a formal process that bidders are invited to attend. The bid prices are read out and bids are examined for compliance with the requirements of the solicitation documents.

Bid Security

(see also: Bid Bond)
A monetary assurance guaranteed by a bank, or other approved institution, which gives the purchaser the right to take the bid security if the bidder (i) withdraws their bid before the end of the bid validity period or, (ii) refuses to sign the contract if selected. The bid security guarantee is determined by the purchaser, including the bank or other institution from which the bid security guarantee may be issued.

Bid Securing Declaration

A non-monetary form of bid security. It is a notarized statement made by a bidder committing to sign the contract if they are selected before the end of the bid validity period stipulated in the solicitation documents. In this sworn statement, the bidder agrees to be automatically disqualified from bidding for any future government contracts for a stipulated period if they either withdraw their bid, fail to sign the contract before the end of the bid validity period or are unable to provide a performance guarantee, if required.

Bid Validity

(see also: Proposal Validity)

A certification from a bidder of the length of time their offer will be valid. At the end of this period, the bidder may grant an extension of the bid validity period, withdraw their bid if the contract has not been signed, or refuse to sign the contract.

Bid Validity Period

(see also: Proposal Validity Period)

The period within which a bidder's offer is considered legally binding. At the end of this period, the bidder may grant an extension of the bid validity period, withdraw their bid if the contract has not been signed, or refuse to sign the contract.

Bill of Quantities (BOQ)

A document showing a list of the units, quantities, rates and amounts of materials, parts, and labor needed to build, maintain, or repair a specific infrastructure.

Buyer

(same as Purchaser)

Depending on the context within which it is used, "buyer" may refer to an individual making purchases on behalf of an organization. It can also refer to the organization itself that is responsible for the purchases. For example: a procuring entity is essentially a buyer, and the procurement specialists working for the procuring entity are also considered buyers.

C

Civil Works Procurement   This is one of the main public procurement categories besides goods and services. Its primary characteristic is the construction, renovation or maintenance of infrastructure. The contract is usually undertaken by a firm because a team of individuals is needed to complete the required tasks of building, renovating or maintaining an infrastructure.

Consulting Services   This procurement category focuses on the hiring of firms or individuals to provide services that are primarily intellectual in nature. Studies, training, advice, and technical assistance are examples of consulting services.

Contract Administration   The surveillance, monitoring and reporting on the performance of suppliers, contractors or service providers, from when the contract is signed until it is either terminated or completed, to ensure compliance with the terms and conditions of the contract.

Contract Closeout   The process undertaken to confirm and ensure that all deliverables and payments under a contract have been made and that all the records on file are complete. This action is also taken before a specific procurement record becomes inactive and ready for electronic or physical storage. The storage of completed procurement records should be done preferably after they have been audited.

Contract Completion Date   The date by which all deliverables under a contract must be satisfied. The date agreed is the one reflected in the initial contract or any amendment to the original contract completion date. The successful completion of the contract marks the end of the agreement between the parties, unless a defects liability, guarantee or warranty period is still in effect.

Contract signing   Marks the date of entry into force of a contract, making it legally binding between the parties to the contract

Contract Suspension   Contract suspension is not the same as the suspension of a firm from participation in the selection process for a stipulated period, usually called debarment. Contract suspension is the cessation of all the activities and performance under a contract resulting from unforeseen events, breach of the contract terms and conditions, or mutual agreement between the parties to the contract. This cessation may be temporary, leading to the resumption of contract execution towards contract completion or may lead to the eventual termination of the contract for default, convenience or Force Majeure.

Contract Termination   The permanent cessation of work under a contract, resulting from unforeseen events, breach of the contract terms and conditions, or mutual agreement between the parties to the contract.

D

Debriefing Report   An oral or written report to bidders that were not successful during the selection process. Depending on the procurement rules, this report is usually prepared at the bidder's request and after the contract is awarded (signed). It should inform the bidders of the results of the evaluation of their bids or proposals and specify the strengths and weaknesses of the same.

Defects Liability   The legal responsibility of a contractor to repair or cover the cost of repair of any defects found in a completed or renovated infrastructure works. This responsibility is only for the period stipulated in the contract, usually 12 months.

Defects Liability Guarantee
A sum of money, usually 10% of the contract value, that is retained from the contractor by the purchaser; it may be replaced with a bank guarantee for the same amount. The purchaser has the right to use the guarantee if the contractor fails to repair any defects found in the completed or renovated infrastructure, before the defects liability period ends.

Defects Liability Period
(also called Defects Notification Period)
The timeframe within which the contractor of a completed or renovated infrastructure works has the responsibility to repair or cover the cost of any defects discovered in the infrastructure. This period is specified in the contract and is usually one year. The defects liability guarantee is retained by the purchaser for the duration of the defects liability period.

E

Entity
Used to denote a department or unit of an organization; e.g. procurement entity, requesting entity, etc.

Evaluation Committee, Evaluation Panel, Technical Evaluation Panel, Bid Review Panel, etc.
A group of three or more qualified individuals selected to evaluate bids or proposals received in response to a procurement notice. The evaluation is done based on the criteria stated in the solicitation documents and the procurement rules, and the evaluation committee recommends the outcome (award or rejection) as a result of their assessment.

F

Firm or Firms
A non-profit or for-profit business organization.

Force Majeure
A delay or suspension of the implementation of a contract, due to an unforeseen event or condition that is not caused by or could not have been prevented by any of the parties to the contract. A Force Majeure event could lead to the temporary suspension or actual termination of the contract. It is not considered a breach of contract because the event is not under the control of any of the parties.

G

Goods One of the primary procurement categories. Any physical item or a combination of an item with a service, where the item is the main product being procured. An example is the procurement of furniture or vehicles, where there is need for assembling and installing the furniture when received, and for periodic servicing, repairs or maintenance of the vehicles procured. These are categorized as goods because the service element is incidental.

Government Procurement, Public Contracting, Public Sector Procurement, Government Buying. All fall under the category of Public Procurement.

I

Implementing Entity The entity responsible for managing a specific project or project activity. The implementing entity may also be a requesting entity responsible for preparing the activity procurement plan, requesting a specific procurement action, participating in the development of the terms of reference and technical specifications of its requirements, and taking part in the administration of the contract. The implementing entity should be represented on the evaluation panel and participate in contract negotiations.

Inactive Contract Files Records of contracts that have reached their completion date, have been terminated, or cancelled, or the period in which they are to remain active has already elapsed. The period a contract record can remain active or inactive is determined either by the procurement rules or other government regulation. Once contracts reach the end of their active life, they are usually kept in storage for another stipulated period. They may also be stored electronically indefinitely.

Infrastructure Works

(same as Civil works and works)

One of the main procurement categories that has to do with the construction, refurbishment, demolition and rebuilding of infrastructure. This includes buildings, bridges, roads, utilities, and similar structures.

J

Joint Venture

A type of association between bidders where all members of the association are proportionally liable to the Client.

K

Kick-off Meeting

A meeting scheduled to take place soon after the contract is signed and before contract commencement, where the parties to the contract meet to introduce the principal representatives of each side and to discuss details of contract execution, including other issues pertaining to contract deliverables.

L

Letter of Credit

A promise to pay, purchased by the buyer (purchaser) from a bank. An agreement to pay the supplier on behalf of the buyer under the conditions stipulated in the letter of credit. Once the supplier complies with the conditions and presents the agreed documentation, the bank is obliged to release the funds to them as indicated in the letter of credit.

M

Manufacturer

The producer of the goods delivered to the purchaser by the supplier. Sometimes the manufacturer is also the supplier. A manufacturer's authorization should be requested when the supplier is not the manufacturer of the goods they supply. This is especially important for the warranty.

Mobilization

The period between signing the contract and the actual commencement of the work or service. All preparations for the commencement of the contract are done during this period as agreed between the parties to the contract.

O

Offer   An offer is the same as a Bid received in response to an invitation for bids or a Proposal received in response to a request for proposals. A quotation received in response to a request for quotation is also considered an offer.

Organization   A for-profit or non-profit entity, government-owned enterprise or educational institution.

P

Parties   Those involved in the signing of the contract and having rights and responsibilities towards each other. The Parties comprise, on the one hand, the purchaser and on the other the supplier, contractor or service provider.

Performance Guarantee   A sum of money retained from payments to a contractor, supplier or service provider, or given to the purchaser in advance in the form of an agreed bank guarantee. The amount of the performance guarantee is usually 10% of the contract value, and is returned to the contractor, supplier or service provider at the end of the satisfactory completion of the contract. The conditions under which the purchaser may retain the performance guarantee are expressed in the contract and in the provisions of the bank guarantee.

Post-qualification   The process of verifying all statements made or documentation received in the selected bidder's offer in order to determine compliance with legal, financial and technical requirements of the solicitation documents.

Pre-qualification   An assessment made before inviting bids or proposals, of the level of experience and capacity of firms expressing interest in undertaking a contract. Upon conclusion of the pre-qualification process, bids or proposals are solicited only from firms shortlisted as a result of the prequalification evaluation.

Procurement Categories

There are three basic procurement groupings: goods, services and works. Under goods, suppliers provide commodities (supplies and equipment). services are consulting and non-consulting services, and works are for rehabilitation, repair or construction of infrastructure.

Procurement Entity

(same as Procuring Entity)

An appointed public body engaged in and responsible for, purchasing goods, services and works, and for awarding contracts or purchase orders.

Procurement Lead-time

The timeframe from the issuing or publication of the solicitation documents, up to when the contract is signed by the parties.

Procurement Legal and Regulatory Framework

(same as Procurement Rules)

Comprises all laws, regulations, manuals and guidelines that govern the management and practice of public procurement and contract administration.

Procurement Methods

Procedures used to acquire goods, services and works. Some methods are competitive; others, non-competitive. The use of competitive procurement methods is preferred.

Procurement Principles

The fundamental values that govern the management and practice of public procurement. They also guide the conduct of public procurement practitioners and other stakeholders involved in the public procurement process.

Procurement Requisition

A formal request to begin the procurement process in order to fulfill a specific procurement requirement. It is also used to certify budget allocation. The procurement process should not begin without an approved procurement requisition certifying the availability of funds.

Procurement Rules

(see: Procurement Legal and Regulatory Framework)

Refers to the procurement legal and regulatory framework which govern the procurement process.

Procuring Entity (see: Procurement Entity)	Same as Procurement Entity.
Proposal (see: Bid)	Offer submitted by business firms in response to a request for proposals.
Proposal Validity (see: Bid Validity)	The period in which a proposal submitted by a firm, in response to a request for proposals, must remain unchanged. After this period, if the contract has not been awarded, the firm is at liberty to extend or cancel the validity without incurring a penalty. Any attempt by the firm to change its proposal before the end of the proposal validity period is unacceptable and may result in the rejection of their proposal.
Public Procurement	The acquisition of goods, services and works with public funds, through a formal process, to facilitate government operations and provide public goods and services.
Purchaser (see: Buyer)	The party responsible for the payment of the goods, services or works acquired.

R

Requesting Entity (see: Implementing Entity)	The entity for which a specific procurement action is carried out.
Responsiveness	The bidder's compliance with the provisions of the solicitation documents determines the bidder's responsiveness. In order to be responsive, a bidder needs to submit all mandatory and administrative documentation in the manner requested in the solicitation documents.

S

Scope of Work	Describes the work required and that needs to be carried out in order to fulfill the contract requirement. Also specifies the activities/tasks and expected deliverables.

Selected Bidder The Bidder recommended for contract award as a result of the evaluation of bids received in response to an invitation for bids.

Services One of the primary procurement categories that deal with all types of consultant and non-consultant services.

Service Provider Firm or individual engaged by the purchaser to fulfill a service requirement (consultant or non-consultant).

Supplier The entity with which the purchaser agrees to fulfill a requirement for goods or services on a one-off basis or over a definite or indefinite period.

Supplier, Contractor, Service Provider, Vendor, Consultant These refer to the source where goods, services and works are obtained. They can be companies or individuals. Sometimes the terms are used synonymously. Supplier or vendor, however, is usually used for goods procurement, and contractor is used for infrastructure works procurement, but it is also used for service provider. Service provider and consultant are sometimes used synonymously; service provider is used for providers of non-consulting services such as equipment maintenance and repair, cleaning services, transport services, etc. Consultant is used for individuals or firms that provide consulting services.

T

Technical Specifications The technical details of the required goods and related services and works described in such a manner that the contractor, supplier or service provider can easily understand and determine their capability to fulfill the requirement.

Tender Board An entity created by law to oversee the public procurement process and to ensure that all public procurement activities are carried out in accordance with the public procurement rules.

Tender Box A receptacle used for receiving tenders (bids/ proposals). The use of the tender box for receiving tender submissions is dependent on the stipulations of the procurement rules. There are two types of tender boxes: electronic and physical.

Tender Documents

(see: Solicitation Documents and Bidding Documents)

The terms tender documents, solicitation documents and bidding documents are referring to the documents which are prepared by the entity needing goods and services. These documents describe what is needed, the specifications or scope of work of what is needed, when it is needed, the qualification requirement and the criteria for evaluation to determine the best qualified bidder. These documents are usually sold but are sometimes available free of charge.

Terms of Reference A document clearly describing the details of a requirement for consultant services. It includes, as a minimum, the background and objectives of the assignment, tasks to be completed, personnel requirement and qualifications, as well as the duration and location of the assignment.

W

Works Procurement

(see: Civil works Procurement)

Same as civil works or infrastructure works procurement.

Suggestions for Further Reading

In addition to the procurement rules and regulations specific to the procurement system you are studying or working with, the following are suggested for further reading and reference:

Public Procurement Regulation: An Introduction, edited by Professor Sue Arrowsmith, University of Nottingham, 2011

UNCITRAL Model Law on Public Procurement, United Nations, New York, 2014

Consulting Services Manual 2006: A Comprehensive Guide to the Selection of Consultants, The World Bank, 2006

Procurement for Local Development: A Guide to Best Practice in Local Government Procurement in Least Developed Countries, UN Capital Development Fund, 2013

Public Procurement Toolbox, OECD

About the Author

This Guide was written by Jorge A. Lynch T., an International Procurement and Supply Management Consultant with almost three decades of experience. He manages, advises, trains and mentors on Public and Project Procurement and has completed assignments in Africa, Asia, the Pacific and Latin America. He is the author of *Public Procurement and Contract Administration: A Brief Introduction*, and received a Master of Science degree in Logistics and Supply Chain Management from Cranfield University in the UK and is a Fellow (FCIPS) of the Chartered Institute of Procurement and Supply (CIPS).

His main interest is to teach public and project procurement management to novice practitioners in order to assist them to develop a solid foundation on which to build a successful career as knowledgeable procurement professionals.

To learn more, you should become a member of the Procurement ClassRoom at: https://procurementclassroom.com/join-free-library/